The Problem Solver™

Student Workbook
Second Edition

Judy Goodnow
Shirley Hoogeboom

www.WrightGroup.com

Wright Group

Copyright © 2008 by Wright Group/McGraw-Hill.

All rights reserved. Except as permitted under the United States Copyright Act, no part of this publication may be reproduced or distributed in any form or by any means, or stored in a database or retrieval system, without the prior written permission from the publisher, unless otherwise indicated.

Printed in the United States of America.

Send all inquiries to:
Wright Group/McGraw-Hill
8787 Orion Place
Columbus, OH 43240

ISBN 978-0-07-704097-0
MHID 0-07-704097-X

6 7 8 9 MAL 13

Recording Sheet

NAME _____

1 FIND OUT

What question do you have to answer?

Find out what the problem tells you.

2 CHOOSE A STRATEGY

I can _____
to solve the problem.

3 SOLVE IT

Show your work.

4 LOOK BACK

- Read the problem again.
- Check your work.
- Did you answer the question?
- Does your answer make sense?
- How do you know?

The Problem Solver 1

Use Logical Reasoning

1

On Ryan's farm there is a rooster, a cat, a rabbit, a dog, and an owl. Ryan likes all the animals, but he likes one best of all.

- It has 4 legs.
- It has ears that stand straight up.
- It has a short tail.

Which animal does Ryan like best?

① FIND OUT

What question do you have to answer?

_____

_____

Find out what the problem tells you.

② CHOOSE A STRATEGY

I can _____ to solve the problem.

3 SOLVE IT

Cross out the animals that don't fit the clues.

Which animal does Ryan like best? _____

4 LOOK BACK

Read the problem again. Check your work.

Use Logical Reasoning

2

Maricela and Jenna are playing a game called Guess My Shape. Jenna draws some shapes and gives Maricela three clues about her secret shape.

- **It has more than three corners.**
- **All of its sides are the same length.**
- **It doesn't have square corners.**

Which shape is Jenna's secret shape?

① FIND OUT

What question do you have to answer?

_____

_____

Find out what the problem tells you.

② CHOOSE A STRATEGY

I can _____ to solve the problem.

4   The Problem Solver

③ SOLVE IT

Cross out the shapes that don't fit the clues.

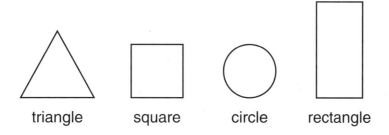

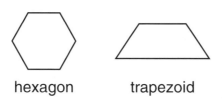

Which shape is Jenna's secret shape? _____

④ LOOK BACK

Read the problem again. Check your work.

Use Logical Reasoning

3

Dan and Delia are going to buy a toy.
They both like one of the toys very much.
Dan and Delia want to buy this toy.

- It has **4 wheels**.
- It does not have a **ladder**.
- It has a **back seat**.

Which toy do Dan and Delia want to buy?

1 FIND OUT

What question do you have to answer?

_____

_____

Find out what the problem tells you.

2 CHOOSE A STRATEGY

I can _____ to solve the problem.

6   The Problem Solver

3 SOLVE IT

Cross out the toys that don't fit the clues.

Which toy do Dan and Delia want to buy? _____

4 LOOK BACK

Read the problem again. Check your work.

The Problem Solver

Make an Organized List

 Crayons

4

Katie Kangaroo is going to school! She takes out her sneakers and socks. Katie has a pair of red sneakers and a pair of yellow sneakers. She has one pair of blue socks and one pair of green socks. What are the different sets of sneakers and socks that Katie can put on today?

1 FIND OUT

What question do you have to answer?

_____

_____

Find out what the problem tells you.

2 CHOOSE A STRATEGY

I can _____ to solve the problem.

8   The Problem Solver

3 SOLVE IT

Color to show the 4 different sets.

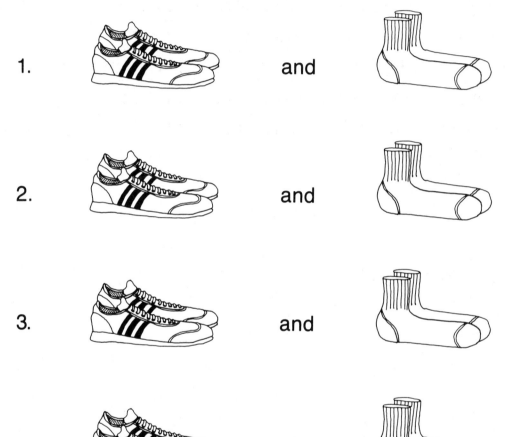

1. and
2. and
3. and
4. and

4 LOOK BACK

Read the problem again. Check your work.

Make an Organized List

 Crayons

5

Tracy's friends are calling to her from outside. They want her to come out and make a snowman. She goes to her closet to get mittens and a cap. She has one pair of striped mittens and one pair of mittens with a diamond pattern. She has one red cap and one blue cap. What are the different sets of snow things that Tracy can wear?

1 FIND OUT

What question do you have to answer to solve the problem?

_____

_____

Find out what the problem tells you.

2 CHOOSE A STRATEGY

I can _____ to solve the problem.

10   The Problem Solver

3 SOLVE IT

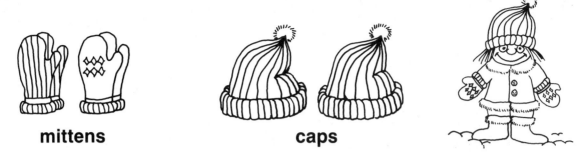

mittens   caps

Draw and color to show the 4 different sets.

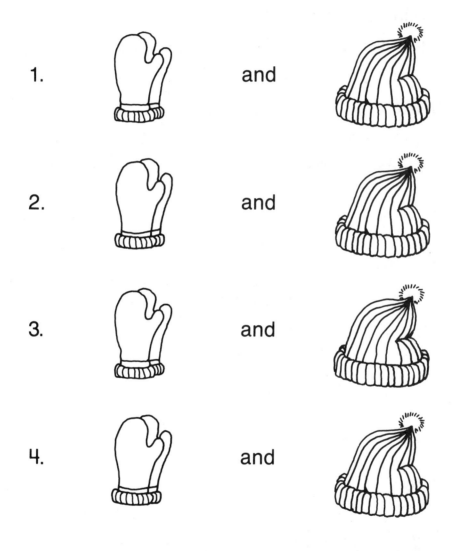

1. [mitten] and [cap]

2. [mitten] and [cap]

3. [mitten] and [cap]

4. [mitten] and [cap]

4 LOOK BACK

Read the problem again. Check your work.

The Problem Solver   11

Make an Organized List

6

Yum! Harvey makes very good sandwiches. Harvey has two sandwich spreads in his food box. They are cheese and peanut butter. He has apple bread and onion bread in his food box. When Harvey makes a sandwich, he puts one sandwich spread on one kind of bread. What are the different sandwiches Harvey can make?

1 FIND OUT

What question do you have to answer to solve the problem?

_____

_____

Find out what the problem tells you.

2 CHOOSE A STRATEGY

I can _____ to solve the problem.

3 SOLVE IT

apple   **onion**

Write the names on the pictures to show the 4 different sandwiches Harvey can make.

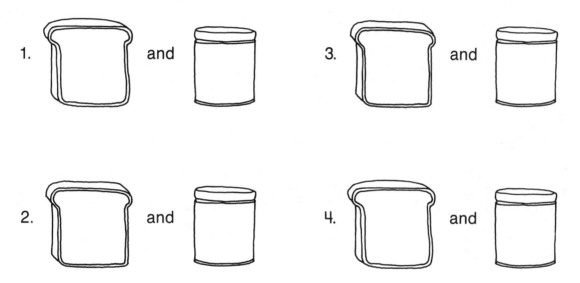

1. ⬜ and ⬜   3. ⬜ and ⬜

2. ⬜ and ⬜   4. ⬜ and ⬜

4 LOOK BACK

Read the problem again. Check your work.

Use or Make a Table

7

Krista is trying to count the goldfish, but they keep hiding under rocks. Lupe gives some clues about how many goldfish there are. Here are Lupe's clues.

- There are more than 5.
- There are fewer than 8.
- There are not 6.

How many goldfish are in the fish bowl?

1 FIND OUT

What question do you have to answer?

_____

_____

Find out what the problem tells you.

2 CHOOSE A STRATEGY

I can _____ to solve the problem.

14   The Problem Solver

3 SOLVE IT

Cross out numbers in the table that don't fit the clues.

2 🐟 🐟

3 🐟 🐟 🐟

4 🐟 🐟 🐟 🐟

5 🐟 🐟 🐟 🐟 🐟

6 🐟 🐟 🐟 🐟 🐟 🐟

7 🐟 🐟 🐟 🐟 🐟 🐟 🐟

8 🐟 🐟 🐟 🐟 🐟 🐟 🐟 🐟

How many goldfish are in the fish bowl? _____

4 LOOK BACK

Read the problem again. Check your work.

Use or Make a Table

8

Aisha's cat had kittens. Tomás wants to count the kittens, but they keep hiding in the barn. Aisha gives Tomás clues about how many kittens there are. Here are Aisha's clues.

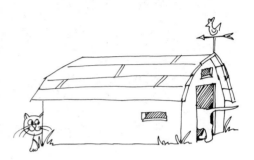

- There are fewer than 10.
- There are more than 7.
- There are not 8.

How many kittens did Aisha's cat have?

❶ FIND OUT

What question do you have to answer?

_____

_____

Find out what the problem tells you.

❷ CHOOSE A STRATEGY

I can _____ to solve the problem.

3 SOLVE IT

Cross out numbers that don't fit the clues.

1
2
3
4
5
6
7
8
9
10

How many kittens did Aisha's cat have? _____

4 LOOK BACK

Read the problem again. Check your work.

Use or Make a Table

9

Dusty was giving a treasure hunt for puppies. He hid the bones in funny places. Dusty did not tell the puppies how many bones he hid. He just gave them some clues.

- There are more than 9.
- There are fewer than 12.
- There are not 10.

How many bones did Dusty hide?

❶ FIND OUT

What question do you have to answer?

_____

_____

Find out what the problem tells you.

❷ CHOOSE A STRATEGY

I can _____ to solve the problem.

18 The Problem Solver

3 SOLVE IT

Cross out the numbers in the table that don't fit the clues.

1 🦴
2 🦴 🦴
3 🦴 🦴 🦴
4 🦴 🦴 🦴 🦴
5 🦴 🦴 🦴 🦴 🦴
6 🦴 🦴 🦴 🦴 🦴 🦴
7 🦴 🦴 🦴 🦴 🦴 🦴 🦴
8 🦴 🦴 🦴 🦴 🦴 🦴 🦴 🦴
9 🦴 🦴 🦴 🦴 🦴 🦴 🦴 🦴 🦴
10 🦴 🦴 🦴 🦴 🦴 🦴 🦴 🦴 🦴 🦴
11 🦴 🦴 🦴 🦴 🦴 🦴 🦴 🦴 🦴 🦴 🦴
12 🦴 🦴 🦴 🦴 🦴 🦴 🦴 🦴 🦴 🦴 🦴 🦴

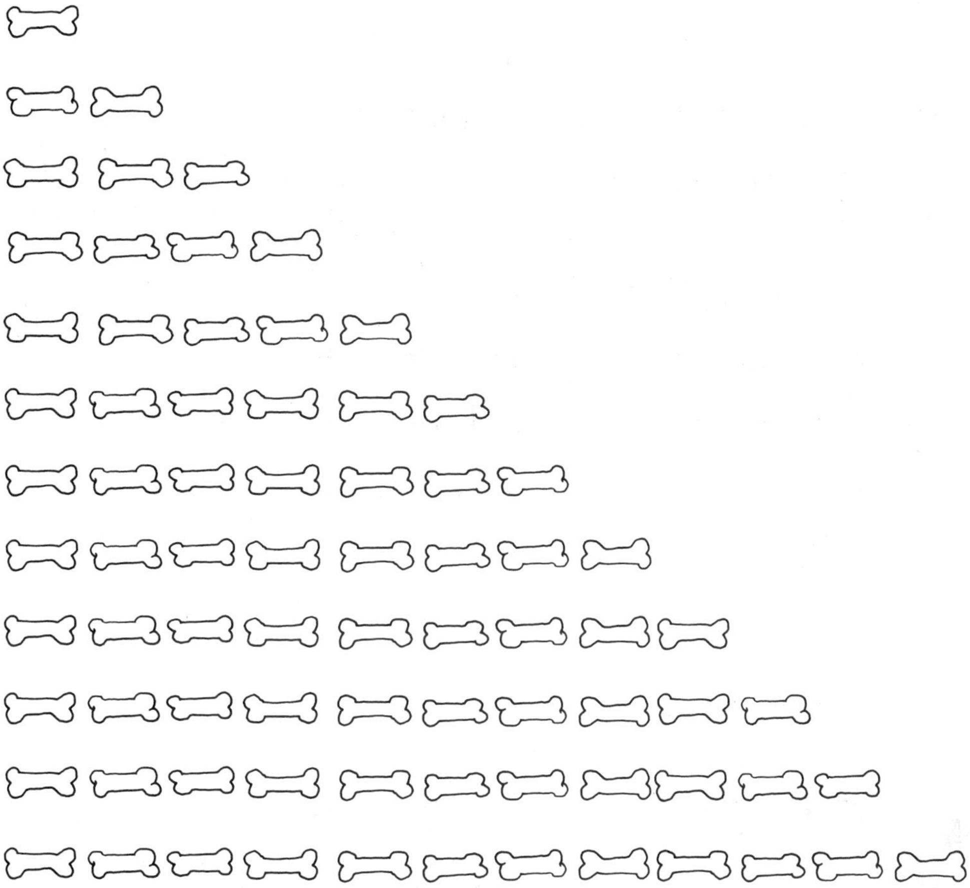

How many bones did Dusty hide? _____

4 LOOK BACK

Read the problem again. Check your work.

Use Logical Reasoning

10

Bob, Peggy, Jeremy, and Doreen are the Hooper children. Each of the children carries things to school in a backpack. Bob's pack is the largest one. Peggy's pack has dots all over it. Doreen's pack does not have a hole in it. Which backpack belongs to each child?

1 FIND OUT

What question do you have to answer?

_____

_____

Find out what the problem tells you.

2 CHOOSE A STRATEGY

I can _____ to solve the problem.

3 SOLVE IT

Draw a line from each child to his or her backpack.

Doreen

Bob

Peggy

Jeremy

4 LOOK BACK

Read the problem again. Check your work.

Use Logical Reasoning

11 David, James, Maria, and Ruth each have a puppet. David's puppet has the largest ears. Ruth's puppet is wearing a hat. Maria's puppet does not have a long nose. Which puppet belongs to each child?

1 FIND OUT

What question do you have to answer to solve the problem?

_____

_____

Find out what the problem tells you.

2 CHOOSE A STRATEGY

I can _____ to solve the problem.

3 SOLVE IT

Draw a line from each child to his or her puppet.

David

James

Maria

Ruth

4 LOOK BACK

Read the problem again. Check your work.

The Problem Solver

Use Logical Reasoning

12

Daisy, Bailey, Elton, and Twig have just finished their baths. Now they need to put their collars back on, but they have gotten all mixed up. The tag on Twig's collar is round. The tag on Bailey's collar is square. The tag on Elton's collar is not a hexagon. Which collar belongs to each dog?

1 FIND OUT

What question do you have to answer to solve the problem?

_____

_____

Find out what the problem tells you.

2 CHOOSE A STRATEGY

I can _____ to solve the problem.

3 SOLVE IT

Draw a line from each dog to his or her collar.

Daisy

Bailey

Elton

Twig

4 LOOK BACK

Read the problem again. Check your work.

The Problem Solver

Use or Make a Picture or Diagram

13

Susan is going to school. She has to cross 8 bridges on her way. She starts out from her house and crosses 6 bridges. Stop! She dropped her lunch somewhere. Susan turns around and goes back across 2 bridges. She finds her lunch. She turns around again and goes across 4 bridges. Where is Susan now?

❶ FIND OUT

What question do you have to answer to solve the problem?

_____

_____

Find out what the problem tells you.

❷ CHOOSE A STRATEGY

I can _____ to solve the problem.

3 SOLVE IT

Write an *S* to show where Susan is.

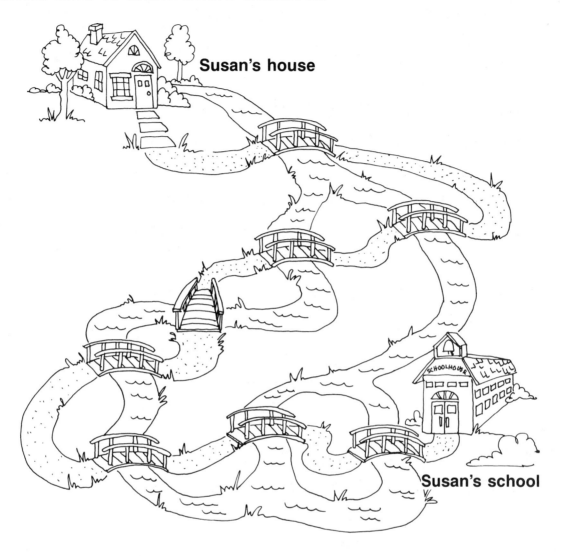

4 LOOK BACK

Read the problem again. Check your work.

Use or Make a Picture or Diagram

14

Barney Bear is going to Bert's birthday party. Barney must pass 8 trees on the way from his home to Bert's home. Barney leaves his home and goes past 6 trees. Oh! Where is his present? Barney dropped it somewhere. He turns around and walks back past 4 trees. There's his present! He takes it and turns around to go toward Bert's home again. He passes 2 trees. Where is Barney now?

① FIND OUT

What question do you have to answer to solve the problem?

_____

_____

Find out what the problem tells you.

② CHOOSE A STRATEGY

I can _____ to solve the problem.

3 SOLVE IT

Write a *B* to show where Barney is.

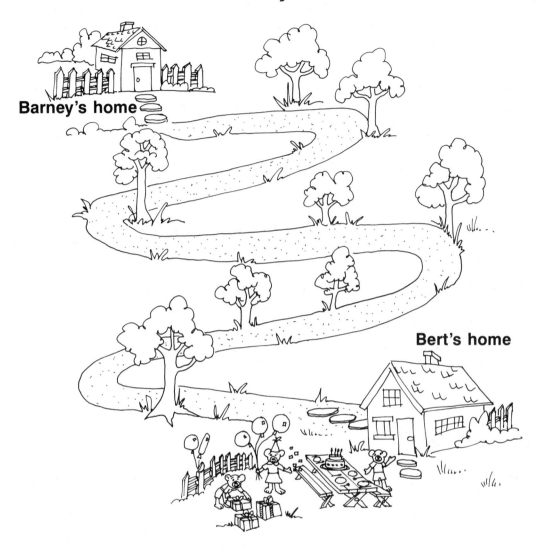

4 LOOK BACK

Read the problem again. Check your work.

Use or Make a Picture or Diagram

15

Rosa is going to a store to get some balloons. She has to pass 12 light posts on the way from her house to the store. Rosa leaves her house and runs past 10 light posts. Oops! She can't find her money. She turns around and goes back past 8 light posts. She finds her money on the sidewalk. Rosa turns around again and runs past 9 light posts. Where is Rosa now?

① FIND OUT

What question do you have to answer to solve the problem?

_____

_____

Find out what the problem tells you.

② CHOOSE A STRATEGY

I can _____ to solve the problem.

3 SOLVE IT

Write an *R* to show where Rosa is.

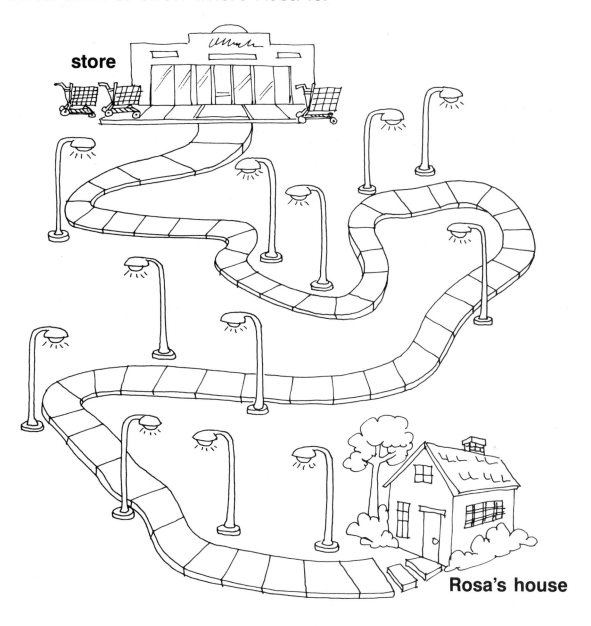

4 LOOK BACK

Read the problem again. Check your work.

Make an Organized List

16

Wally is looking at the yummy food in the lunchroom. He wants to eat 7 pieces of food for lunch today. Wally will take only 2 plates of food. He will eat everything on those plates. What are the 3 different lunches that Wally could buy?

1 FIND OUT

What question do you have to answer?

_____

_____

Find out what the problem tells you.

2 CHOOSE A STRATEGY

I can _____ to solve the problem.

3 SOLVE IT

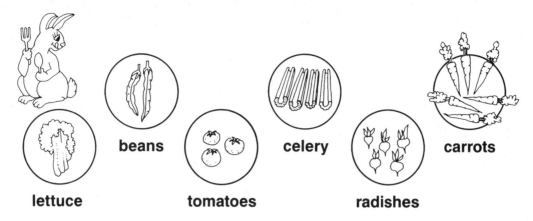

Finish the list to show the answer.

Lunches with 7 pieces of food:

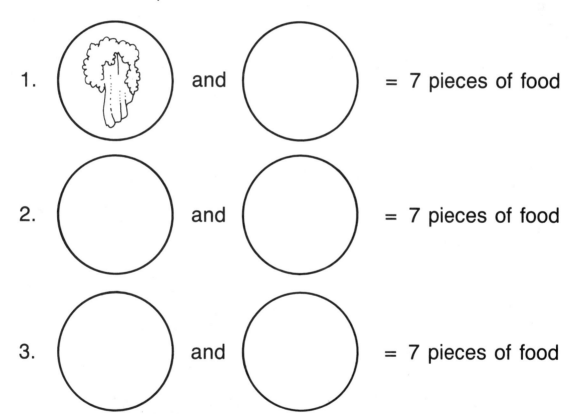

1. (lettuce) and ( ) = 7 pieces of food

2. ( ) and ( ) = 7 pieces of food

3. ( ) and ( ) = 7 pieces of food

4 LOOK BACK

Read the problem again. Check your work.

Make an Organized List

17

Mary is playing a card game with her friends. Mary has cards in her hand. It is her turn to play. She has to lay down 2 cards. The cards must show 9 hearts all together. What 3 different pairs of cards can Mary lay down?

1 FIND OUT

What question do you have to answer to solve the problem?

_____

_____

Find out what the problem tells you.

2 CHOOSE A STRATEGY

I can _____ to solve the problem.

34   The Problem Solver

3 SOLVE IT

Finish the list to show the answer.

Pairs of cards that show 9 hearts:

1.    and   [   ]   = 9 hearts

2. [   ]   and   [   ]   = 9 hearts

3. [   ]   and   [   ]   = 9 hearts

4 LOOK BACK

Read the problem again. Check your work.

Make an Organized List

18

Leo and Benji put 6 tin cans on the fence. Each tin can has a number on it. Leo and Benji take turns throwing stones at the cans. Leo and Benji try to knock down 2 cans on each turn. If the numbers on the 2 cans add up to 10, the player gets 10 points. What are the 3 different pairs of cans a player can knock down to get 10 points?

❶ FIND OUT

What question do you have to answer?

_____

_____

Find out what the problem tells you.

❷ CHOOSE A STRATEGY

I can _____ to solve the problem.

36   The Problem Solver

3 SOLVE IT

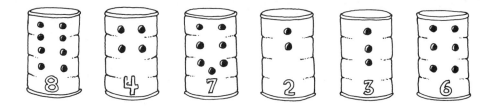

Finish the list to show the answer.

Pairs of cans that give 10 points:

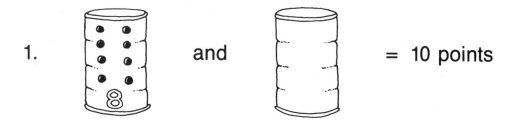

1. [can with 8] and [   ] = 10 points

2. [   ] and [   ] = 10 points

3. [   ] and [   ] = 10 points

4 LOOK BACK

Read the problem again. Check your work.

Guess and Check

19

Juan looked at the toys in Skip's Junk Shop. Skip is selling some little green turtles, rubber birds, squeaky bears, wooden dogs, and fuzzy cats. Juan bought 3 different toys. He paid 12 cents in all. Which 3 toys did Juan buy?

1 FIND OUT

What question do you have to answer to solve the problem?

_____

_____

Find out what the problem tells you.

2 CHOOSE A STRATEGY

I can _____ to solve the problem.

3 SOLVE IT

Guess and check to find the answer.

| turtle | bird | bear | dog | cat |
| 3 cents | 4 cents | 8 cents | 9 cents | 1 cent |

Guess: _____

Guess: _____

Guess: _____

Write the names of the toys Juan bought.

_____   _____   _____

4 LOOK BACK

Read the problem again. Check your work.

Guess and Check

20

"Beth, we can only ride for 10 more minutes," said Boris. "Which rides can we take?" Beth looked at the list of rides. She and Boris had tickets for 3 rides. She looked for 3 rides that would take 10 minutes all together. Which 3 different rides can Beth and Boris take?

1 FIND OUT

What question do you have to answer?

_____

_____

Find out what the problem tells you.

2 CHOOSE A STRATEGY

I can _____ to solve the problem.

40   The Problem Solver

3 SOLVE IT

Guess and check to find the answer.

TEACUP TWIST	6 minutes
GHOST TRAIN	5 minutes
SNAKE	4 minutes
GRASSHOPPER	3 minutes
GOOSE BUMP	2 minutes

Guess: _____

Guess: _____

Guess: _____

Write the names of the 3 rides that Beth and Boris can take.

_____   _____   _____

4 LOOK BACK

Read the problem again. Check your work.

The Problem Solver 41

Guess and Check

21

Patrick wants to make money to buy a special gift for his mom. He needs 8 dollars all together. Patrick's neighbors will pay him to do jobs for them. On Saturday, he did 3 different jobs and earned exactly the right amount! Which 3 jobs did Patrick do?

1 FIND OUT

What question do you have to answer to solve the problem?

_____

_____

Find out what the problem tells you.

2 CHOOSE A STRATEGY

I can _____ to solve the problem.

42 The Problem Solver

3 SOLVE IT

Guess and check to find the answer.

Rake
$4

Walk the dog
$2

Wash the car
$5

Water flowers
$3

Put out garbage
$1

Guess: _____

Guess: _____

Guess: _____

Write the names of the jobs that Patrick did.

_____ _____ _____

4 LOOK BACK

Read the problem again. Check your work.

The Problem Solver   43

Use or Look for a Pattern

22

When Bert blows bubbles, everyone watches. No one else can blow bubbles like he can. Bert blows round bubbles and square bubbles, and he blows them in a pattern! Look at Bert's bubbles and find his pattern. What kind of bubble will Bert blow next?

1 FIND OUT

What question do you have to answer to solve the problem?

_____

_____

Find out what the problem tells you.

2 CHOOSE A STRATEGY

I can _____ to solve the problem.

44   The Problem Solver

3 SOLVE IT

Draw the bubble to show the answer.

4 LOOK BACK

Read the problem again. Check your work.

Use or Look for a Pattern

23 Mr. Wing lined up his owls and parrots. They are going to have their picture taken. Mr. Wing used a pattern. Look for his pattern in the line of birds. What will Mr. Wing put in line next, an owl or a parrot?

1 FIND OUT

What question do you have to answer to solve the problem?

_____

_____

Find out what the problem tells you.

2 CHOOSE A STRATEGY

I can _____ to solve the problem.

3 SOLVE IT

Write the name of the bird on the line.

owl **parrot**

4 LOOK BACK

Read the problem again. Check your work.

Crayons

Use or Look for a Pattern

24

Grace Goose is making a string of beads to go around her neck. She is putting beads on the string in a pattern. She is using red beads with circles, green beads with ovals, and blue beads with triangles. Look for Grace's pattern. What kind of bead will Grace put on the string next?

1 FIND OUT

What question do you have to answer to solve the problem?

_____

_____

Find out what the problem tells you.

2 CHOOSE A STRATEGY

I can _____ to solve the problem.

3 SOLVE IT

Draw the bead on the string and color it.

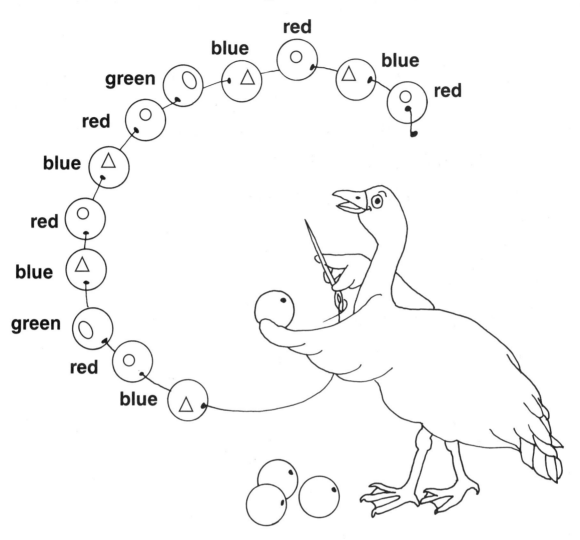

4 LOOK BACK

Read the problem again. Check your work.

Act Out or Use Objects

 Crayons and counters

25

Four fish line up to swim through the castle in their fish tank. The red fish is first in line. The blue fish is in front of the yellow fish. The green fish is in front of the blue fish. What color is the last fish in the line?

1 FIND OUT

What question do you have to answer to solve the problem?

_____

_____

Find out what the problem tells you.

2 CHOOSE A STRATEGY

I can _____ to solve the problem.

3) SOLVE IT

Color the fish to show the answer.

What color is the last fish in the line? _____

4) LOOK BACK

Read the problem again. Check your work.

Paper

Act Out or Use Objects

26 Four boys lined up to go on the Twirl-a-Whirl. Bob was the last one in line. Sam was waiting behind Tom. Pat was behind Sam. Who was first in line?

❶ FIND OUT

What question do you have to answer to solve the problem?

_____

_____

Find out what the problem tells you.

❷ CHOOSE A STRATEGY

I can _____ to solve the problem.

52 The Problem Solver

3 SOLVE IT

Write each boy's name on his shirt.

Who was first in line? _____

4 LOOK BACK

Read the problem again. Check your work.

Paper

Act Out or Use Objects

27

Four animals live in a tree. The owl lives at the top of the tree. The raccoon lives below the lizard. The lizard lives below the squirrel. Where does each animal live in the tree?

1 FIND OUT

What question do you have to answer to solve the problem?

_____

_____

Find out what the problem tells you.

2 CHOOSE A STRATEGY

I can _____ to solve this problem.

54   The Problem Solver

3 SOLVE IT

Write each animal's name in its place on the tree.

4 LOOK BACK

Read the problem again. Check your work.

Use or Make a Table

28

Rose and Rocky Raccoon are going to have a party. They wrote notes to their friends to ask them to come. Rocky wrote faster than Rose did. In the time it took Rose to write 1 note, Rocky wrote 3 notes. Rocky and Rose kept writing in the same way until they finished all the notes. Then Rose put stamps on the 4 notes she wrote. How many notes did Rocky write?

1 FIND OUT

What question do you have to answer?

_____

_____

Find out what the problem tells you.

2 CHOOSE A STRATEGY

I can _____ to solve the problem.

3 SOLVE IT

Finish the table.

Number of Notes Rose Wrote	1	2	3	4
Number of Notes Rocky Wrote	☐☐☐ 3	6		

How many notes did Rocky write? _____

4 LOOK BACK

Read the problem again. Check your work.

Use or Make a Table

29

Millie and Minnie had a race to see who could get the most crumbs. Millie was much faster than Minnie. In the time it took Minnie to get 1 crumb, Millie picked up 4 crumbs. Millie and Minnie kept picking up crumbs in the same way until they felt footsteps shake the ground. Minnie picked up 4 crumbs in all. How many crumbs did Millie get?

1 FIND OUT

What question do you have to answer?

_____

_____

Find out what the problem tells you.

2 CHOOSE A STRATEGY

I can _____ to solve the problem.

3 SOLVE IT

Finish the table.

Number of Crumbs Minnie Got	1	2	3	4
Number of Crumbs Millie Got	○○○○ 4			

How many crumbs did Millie get? _____

4 LOOK BACK

Read the problem again. Check your work.

Use or Make a Table

30

Mark and Jon blew up balloons for a party. It took Jon a long time to blow up a balloon. In the time it took Jon to blow up 1 balloon, Mark blew up 2 balloons. The boys kept blowing up balloons in the same way until all the balloons were done. Jon finished 7 balloons. How many balloons did Mark blow up?

① FIND OUT

What question do you have to answer to solve the problem?

_____

_____

Find out what the problem tells you.

② CHOOSE A STRATEGY

I can _____ to solve the problem.

3 SOLVE IT

Finish the table.

Number of Balloons Jon Blew Up	1	2	3	4	5	6	7
Number of Balloons Mark Blew Up	○○ 2						

How many balloons did Mark blow up? _____

4 LOOK BACK

Read the problem again. Check your work.

The Problem Solver 61

Use or Look for a Pattern

Counters

31

Lolly likes the capital letter L at the beginning of her name. She likes to make the letter L with blocks. She made her first L with 4 blocks. She added 2 blocks to her first L to make her second L. She added 2 blocks to her second L to make her third L. Lolly keeps using her number pattern. How many blocks will Lolly use to make her fifth L?

1 FIND OUT

What question do you have to answer to solve the problem?

_____

_____

Find out what the problem tells you.

2 CHOOSE A STRATEGY

I can _____ to solve the problem.

3 SOLVE IT

Finish the table.

Number of Lolly's L	Number of Blocks in L
1st	4
2nd	6
3rd	8
4th	
5th	

1st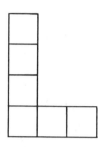

2nd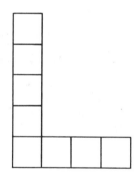

3rd

How many blocks will Lolly use to make her fifth L? _____

4 LOOK BACK

Read the problem again. Check your work.

Use or Look for a Pattern

32 Beth is counting the fruit on the plum tree in her back yard. The first day, Beth counted 5 plums. The next day, there were 3 new plums. Each day, there were 3 new plums. How many plums in all were on the plum tree on the sixth day?

1 FIND OUT

What question do you have to answer to solve the problem?

_____

_____

Find out what the problem tells you.

2 CHOOSE A STRATEGY

I can _____ to solve the problem.

3 SOLVE IT

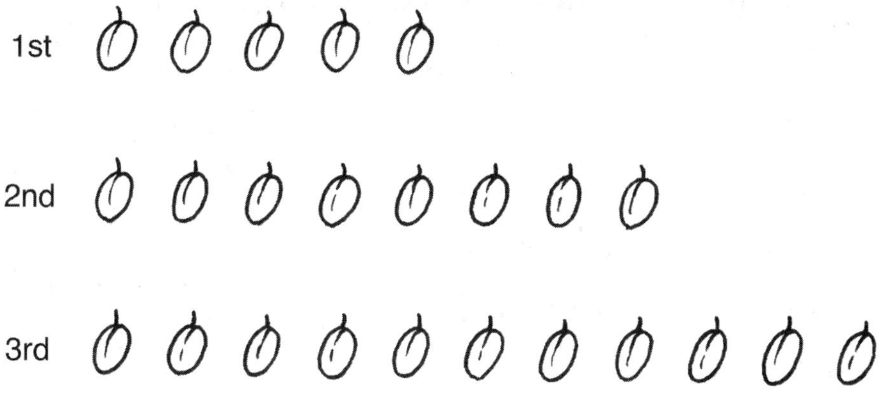

1st O O O O O

2nd O O O O O O O O

3rd O O O O O O O O O O O

Finish the table.

Day	Number of Plums
1st	5
2nd	8
3rd	
4th	
5th	
6th	

How many plums were on the tree on the sixth day? _____

4 LOOK BACK

Read the problem again. Check your work.

The Problem Solver

Use or Look for a Pattern

33

The frogs had a jumping contest. They jumped over a row of stones. They put 3 stones in the row for game 1. They added 3 more stones to the row for game 2. They added 3 more stones to the row for game 3. The frogs kept using the same number pattern to add stones to the row. How many stones were in the row for game 6?

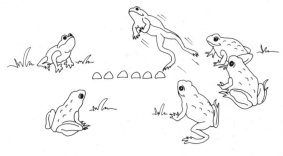

❶ FIND OUT

What question do you have to answer?

_____

_____

Find out what the problem tells you.

❷ CHOOSE A STRATEGY

I can _____ to solve the problem.

3 SOLVE IT

Finish the table.

Game	Number of Stones in the Row
1st	3
2nd	6
3rd	
4th	
5th	
6th	

Game 1 ○ ○ ○

Game 2 ○ ○ ○ ○ ○ ○

Game 3 ○ ○ ○ ○ ○ ○ ○ ○ ○

How many stones were in the row for game 6? _____

4 LOOK BACK

Read the problem again. Check your work.

The Problem Solver 67

Act Out or Use Objects

 Counters

34

The Pet Café is just for pets. They come from all over the neighborhood to enjoy special treats. Today there are 8 pets at the café. They are sitting at 3 tables. There are 5 pets in all at table A and table B. There are 7 pets in all at table B and table C. How many pets are at each table?

1 FIND OUT

What question do you have to answer to solve the problem?

_____

_____

Find out what the problem tells you.

2 CHOOSE A STRATEGY

I can _____ to solve the problem.

68   The Problem Solver

3 SOLVE IT

Put counters on the tables to show the pets.

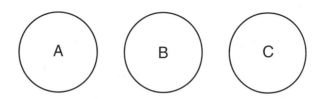

How many pets are at each table?

A _____   B _____   C _____

4 LOOK BACK

Read the problem again. Check your work.

Act Out or Use Objects

Counters

35

Skyler is making a dozen mud pies. They are oh, so ooey gooey! Skyler puts the mud pies on 3 plates to bake in the sun. There are 6 mud pies in all on plate A and plate B. There are 8 pies in all on plate B and plate C. How many mud pies are on each plate?

1 FIND OUT

What question do you have to answer to solve the problem?

_____

_____

Find out what the problem tells you.

2 CHOOSE A STRATEGY

I can _____ to solve the problem.

70 The Problem Solver

3 SOLVE IT

Put counters on the plates to show the mud pies.

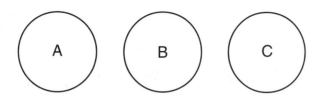

How many mud pies are on each plate?

A _____   B _____   C _____

4 LOOK BACK

Read the problem again. Check your work.

Act Out or Use Objects

 Counters

36

Steamy Swamp is the perfect place for frogs. There are frogs on logs all over the swamp! There are 11 frogs sunning on 3 logs. There are 9 frogs in all on log A and log B. There are 6 frogs in all on log B and log C. How many frogs are on each log?

1 FIND OUT

What question do you have to answer to solve the problem?

_____

_____

Find out what the problem tells you.

2 CHOOSE A STRATEGY

I can _____ to solve the problem.

3 SOLVE IT

Put cubes on the logs to show the frogs.

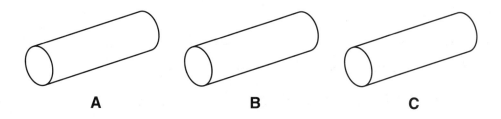

How many frogs are on each log?

A _____ B _____ C _____

4 LOOK BACK

Read the problem again. Check your work.

Use or Make a Table

37

Chuck and Dino rode in a bike parade. Chuck led the parade on his 2-wheel bike. Dino rode on his 3-wheel bike behind Chuck's bike. The rest of the monkeys rode 3-wheel bikes behind Dino's bike. There were 5 bikes all together in the parade. Chuck and Dino put streamers on every wheel in the parade. How many wheels were in the parade?

1 FIND OUT

What question do you have to answer to solve the problem?

_____

_____

Find out what the problem tells you.

2 CHOOSE A STRATEGY

I can _____ to solve the problem.

3 SOLVE IT

Finish the table.

Number of Bikes in the Parade	1	2	3	4	5
Number of Wheels in the Parade	2	5			

How many wheels were in the parade? _____

4 LOOK BACK

Read the problem again. Check your work.

Use or Make a Table

38

It was summertime and the sun was hot. Dawn opened a lemonade stand. She sold lemonade with ice in it. Her friend Carl came to buy some lemonade. Carl paid 4 cents for his first glass of lemonade and 2 cents for each glass of lemonade after that. Carl kept drinking lemonade. At two o'clock, Carl drank his fifth glass of lemonade. How much money did Carl pay Dawn for all the lemonade he drank?

1 FIND OUT

What question do you have to answer to solve the problem?

_____

_____

Find out what the problem tells you.

2 CHOOSE A STRATEGY

I can _____ to solve the problem.

3 SOLVE IT

Finish the table.

Number of Glasses of Lemonade	1	2	3	4	5
How Much Money Carl Paid in All	4¢	6¢			

How much money did Carl pay Dawn? _____

4 LOOK BACK

Read the problem again. Check your work.

Use or Make a Table

39

"It's time for the Pig Picnic!" called out the 4 pigs in the Oink family. Other pig families hurried over to join in the fun. The Curly family came first. Then the Snuffle family, the Snort family, and the Snout family came to the Pig Picnic. There were 3 pigs in each of those families. There were 5 pig families in all at the picnic. They ate lots of corn! How many pigs were there in all at the picnic?

1 FIND OUT

What question do you have to answer to solve the problem?

_____

_____

Find out what the problem tells you.

2 CHOOSE A STRATEGY

I can _____ to solve the problem.

3 SOLVE IT

Finish the table.

Number of Families	1	2	3	4	5
Number of Pigs in All	4	7			

How many pigs were there in all at the picnic? _____

4 LOOK BACK

Read the problem again. Check your work.

Use or Look for a Pattern

Counters

40

Simon Squirrel's cousins are coming to visit him next week. He has been getting food ready for their visit. On Monday he cracked 1 nut. On Tuesday he cracked 3 nuts. On Wednesday he cracked 5 nuts, and on Thursday he cracked 7 nuts. Simon is following a number pattern, so he knows how many nuts he will crack on Friday. How many nuts will he crack on Friday?

1 FIND OUT

What question do you have to answer?

_____

_____

Find out what the problem tells you.

2 CHOOSE A STRATEGY

I can _____ to solve the problem.

80   The Problem Solver

3 SOLVE IT

Fill in the table.

Monday	Tuesday	Wednesday	Thursday	Friday

How many nuts will Simon crack on Friday? _____

4 LOOK BACK

Read the problem again. Check your work.

Use or Look for a Pattern

41

There are lots of tadpoles in the small pond behind Trudy's house. Trudy put her net into the pond, and 3 tadpoles swam into it. She put her net in again, and 6 tadpoles swam into it. She put her net in again, and 9 tadpoles swam into it. Trudy could see that the tadpoles were following a number pattern. She put her net in again. How many tadpoles swam into Trudy's net the fourth time she put it into the pond?

① FIND OUT

What question do you have to answer?

_____

_____

Find out what the problem tells you.

② CHOOSE A STRATEGY

I can _____ to solve the problem.

3 SOLVE IT

Fill in the table.

1st Time	2nd Time	3rd Time	4th Time

How many tadpoles swam into Trudy's net the fourth time? _____

4 LOOK BACK

Read the problem again. Check your work.

Use or Look for a Pattern

42

There are 12 people standing in line for the swan boat ride. The first boat comes by, and some people get on. Now there are 10 people in line. After the second boat comes by, there are 8 people left in line. The third boat comes by. Now there are 6 people left in line. The number of people in line keeps changing in the same way. How many people are left in line after the fourth boat comes by?

1 FIND OUT

What question do you have to answer?

_____

_____

Find out what the problem tells you.

2 CHOOSE A STRATEGY

I can _____ to solve the problem.

3 SOLVE IT

Fill in the table.

At First	After 1st Boat	After 2nd Boat	After 3rd Boat	After 4th Boat

How many people are left in line after the fourth boat comes by? _____

4 LOOK BACK

Read the problem again. Check your work.

Work Backwards

Counters

43

Herman and Polly say Frank catches the most fish because he has the best worms. Maybe they are right. Frank Fox does catch fish! Today he caught 5 more fish than Polly caught. Polly Pig pulled up 4 more fish than Herman did. Herman Rabbit caught only 2 fish. How many fish did Frank catch?

1 FIND OUT

What question do you have to answer to solve the problem?

_____

_____

Find out what the problem tells you.

2 CHOOSE A STRATEGY

I can _____ to solve the problem.

3 SOLVE IT

Use counters to show the fish.

Frank	Polly	Herman

How many fish did Frank catch? _____

4 LOOK BACK

Read the problem again. Check your work.

Work Backwards

Counters

44

Winter is coming. It's time for the chipmunks to hunt for berries. They will put the berries into their storehouse for winter. Lisa found 5 more berries than Jared did today. Jared found 6 more berries than Tim did. Tim stopped to play, so he brought home just 2 berries. How many berries did Lisa find?

1 FIND OUT

What question do you have to answer to solve the problem?

_____

_____

Find out what the problem tells you.

2 CHOOSE A STRATEGY

I can _____ to solve the problem.

88   The Problem Solver

3 SOLVE IT

Use counters to show the berries.

Lisa	Jared	Tim

How many berries did Lisa find? _____

Tim

Jared **Lisa**

4 LOOK BACK

Read the problem again. Check your work.

Work Backwards

45

Seth, Larry, and Todd are collecting Goofy Gumbo cards. Seth has the most cards. He has 7 more cards than Larry has. Larry has 5 more cards than Todd has. Sometimes Todd forgets where he puts his cards, so he only has 3. How many cards does Seth have?

1 FIND OUT

What question do you have to answer to solve the problem?

_____

_____

Find out what the problem tells you.

2 CHOOSE A STRATEGY

I can _____ to solve the problem.

3 SOLVE IT

Larry   Todd   Seth

Use counters to show the cards.

Seth	Larry	Todd

How many cards does Seth have? _____

4 LOOK BACK

Read the problem again. Check your work.

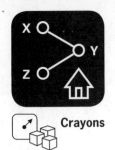

Use or Make a Picture or Diagram

Crayons

46

Cheery Mouse lives in an old tree trunk. Cheery can choose from 2 tunnels to get into her house. Both tunnels go to Cheery's living room. Then there are 3 tunnels that go from the living room to her kitchen. Cheery goes to her kitchen as soon as she comes home. She likes to take a different path every time. What are the 6 different paths Cheery can take from the outside to her kitchen?

1 FIND OUT

What question do you have to answer to solve the problem?

_____

_____

Find out what the problem tells you.

2 CHOOSE A STRATEGY

I can _____ to solve the problem.

92   The Problem Solver

3 SOLVE IT

Use a different color to trace each path.

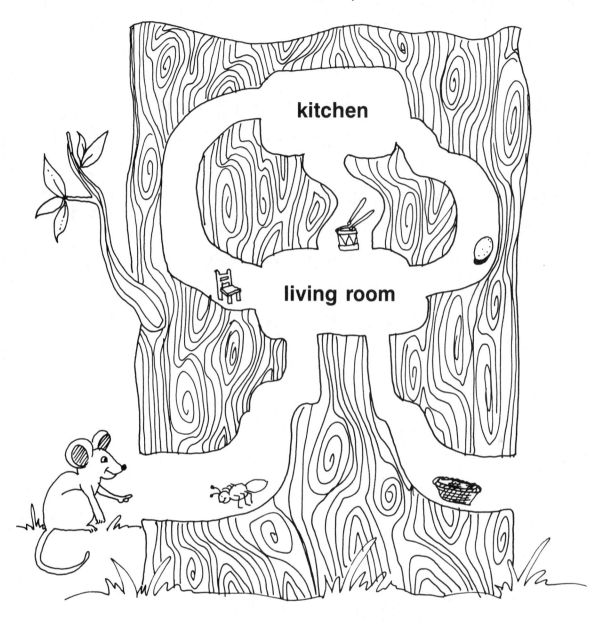

4 LOOK BACK

Read the problem again. Check your work.

The Problem Solver 93

Use or Make a Picture or Diagram

Crayons

47

Gregory Groundhog lives in a cozy home under the ground. Gregory has 3 holes that lead underground. Each hole takes Gregory into a big tunnel that has 2 doors. Gregory can go through either door to get into his home. What are the 6 different paths Gregory can take from above the ground down into his home?

1 FIND OUT

What question do you have to answer to solve the problem?

_____

_____

Find out what the problem tells you.

2 CHOOSE A STRATEGY

I can _____ to solve the problem.

3 SOLVE IT

Use a different color to trace each path.

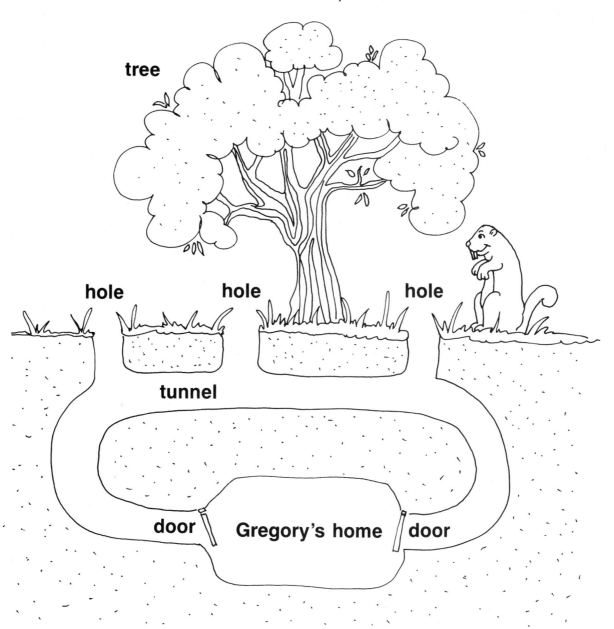

4 LOOK BACK

Read the problem again. Check your work.

The Problem Solver

Use or Make a Picture or Diagram

 Crayons

48

Mr. Small loves his little house. He has a special room with a desk where he likes to work. There are two doors to his house from outside. Inside, there are 3 different entrances to the room with his desk. Mr. Small likes to take a different path to his desk each time he comes in. What are all the different paths Mr. Small can take from outside his house to his desk?

1 FIND OUT

What question do you have to answer to solve the problem?

_____

_____

Find out what the problem tells you.

2 CHOOSE A STRATEGY

I can _____ to solve the problem.

96   The Problem Solver

3 SOLVE IT

Use a different color to trace each path.

4 LOOK BACK

Read the problem again. Check your work.

The Problem Solver

Make It Simpler

 Ribbons or strings

49

Ann, Betsy, Carol, and Dee are taking swimming lessons at the park. They are getting ready to give a show. One part of the show is very pretty. The girls hold colored ribbons and float in the water. Each girl is joined to each of the other girls by a ribbon. How many ribbons do the girls need?

1 FIND OUT

What question do you have to answer to solve the problem?

_____

_____

Find out what the problem tells you.

2 CHOOSE A STRATEGY

I can _____ to solve the problem.

98   The Problem Solver

3 SOLVE IT

Write the numbers.

How many ribbons do the girls need if

2 swimmers are swimming in the show? _____

3 swimmers are swimming in the show? _____

4 swimmers are swimming in the show? _____

How many ribbons do the girls need?

4 LOOK BACK

Read the problem again. Check your work.

The Problem Solver 99

Make It Simpler

50 Milo, June, Andrew, and Joan meet at the park on Saturday mornings. When they meet, each friend slaps hands with each of the other friends. How many slaps are there in all when the friends meet?

1 FIND OUT

What question do you have to answer to solve the problem?

_____

_____

Find out what the problem tells you.

2 CHOOSE A STRATEGY

I can _____ to solve the problem.

3 SOLVE IT

Write the numbers.

How many slaps are there if

2 friends meet at the park? _____

3 friends meet at the park? _____

4 friends meet at the park? _____

How many hand slaps are there in all? _____

4 LOOK BACK

Read the problem again. Check your work.

Make It Simpler

51

Marty, Gino, Heather, Latanya, and Pam are in Mr. Mustard's class. They are having a checkers tournament. Mr. Mustard asked everyone to play each of the other children once. Then the person who had won the most games would be the checker champion. So each child played each other child in one game. How many different games of checkers were played all together?

1 FIND OUT

What question do you have to answer to solve the problem?

_____

_____

Find out what the problem tells you.

2 CHOOSE A STRATEGY

I can _____ to solve the problem.

3 SOLVE IT

Write the numbers.

How many different games are there if

2 children are in the tournament? _____

3 children are in the tournament? _____

4 children are in the tournament? _____

5 children are in the tournament? _____

How many games were played in all? _____

4 LOOK BACK

Read the problem again. Check your work.

Brainstorm

 Counters

52

Fiona is giving you a puzzle to solve. She says, "There are 2 circles on the floor. One circle is large. The other circle is small. There are 3 lions in the large circle. There are 2 lions in the small circle. But there are only 3 lions in all. Where is the small circle, and where are the lions?"

1 FIND OUT

What question do you have to answer to solve the problem?

_____

_____

Find out what the problem tells you.

2 CHOOSE A STRATEGY

I can _____ to solve the problem.

3 SOLVE IT

This is the large circle.

Draw the small circle.

4 LOOK BACK

Read the problem again. Check your work.

Counters

Brainstorm

53

Look at this group of 6 baby chicks. It looks like a triangle pointing up to the top of the page.

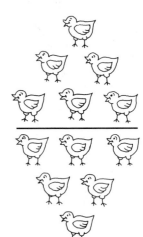

Now look at the chicks again. Only 2 chicks have moved. But now the group looks like a triangle pointing down to the bottom of the page. Which 2 chicks moved?

1 FIND OUT

What question do you have to answer?

_____

Find out what the problem tells you.

2 CHOOSE A STRATEGY

I can _____ to solve the problem.

3 SOLVE IT

In both pictures, color the chicks that moved.

4 LOOK BACK

Read the problem again. Check your work.

 Counters

Brainstorm

54

Marne is giving you a puzzle to solve. He says, "I saw a trail of snails in my garden. There were 2 snails behind a snail and 2 snails in front of a snail. There was a snail between 2 snails, too. What is the smallest number of snails there could be in that trail?"

1 FIND OUT

What question do you have to answer to solve the problem?

_____

_____

Find out what the problem tells you.

2 CHOOSE A STRATEGY

I can _____ to solve the problem.

3 SOLVE IT

Write the number. _____

4 LOOK BACK

Read the problem again. Check your work.

Use or Make a Picture or Diagram

55

Mr. Hall's class voted for their favorite recess games. The graph shows how many children voted for each game.

- Sean's favorite game was the one that most of the boys liked.
- Equal numbers of boys and girls liked Rachel's favorite game.
- Two more girls than boys liked Alyssa's favorite game.

Which games do Sean, Rachel, and Alyssa like best?

1 FIND OUT

What question do you have to answer?

_____

Find out what the problem tells you.

2 CHOOSE A STRATEGY

I can _____ to solve the problem.

3 SOLVE IT

Use the graph and the clues.

Favorite Recess Games

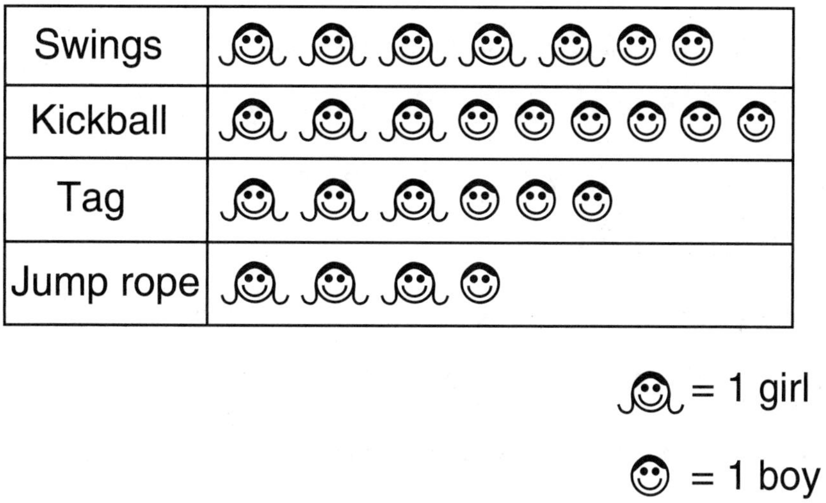

Swings	👧 👧 👧 👧 👧 🙂 🙂
Kickball	👧 👧 👧 🙂 🙂 🙂 🙂 🙂 🙂
Tag	👧 👧 👧 🙂 🙂 🙂
Jump rope	👧 👧 👧 🙂

👧 = 1 girl

🙂 = 1 boy

Which games did Sean, Rachel, and Alyssa vote for?

Sean _____

Rachel _____

Alyssa _____

4 LOOK BACK

Read the problem again. Check your work.

Use or Make a Picture or Diagram

56

One day, the class made a graph to show how each child got to school.

- Isabelle used the way that got two fewer marks than *car*.
- Julian's way was the way most of the children got to school that day.
- Lauren's transportation got more marks than *bus* but fewer than *bike*.

How did Isabelle, Julian, and Lauren get to school?

1 FIND OUT

What question do you have to answer?

_____

_____

Find out what the problem tells you.

2 CHOOSE A STRATEGY

I can _____ to solve the problem.

3 SOLVE IT

Use the graph and the clues.

How We Got to School

Scooter	X	X								
Bike	X	X	X	X	X	X	X			
Bus	X	X	X							
Car	X	X	X	X	X					
Walk	X	X	X	X	X	X	X	X	X	

How did Isabelle, Julian, and Lauren get to school?

Isabelle _____

Julian _____

Lauren _____

4 LOOK BACK

Read the problem again. Check your work.

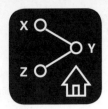

Use or Make a Picture or Diagram

57

There are 6 friends in the Critter Club. Each friend has a different kind of pet. For instance, John has 2 snakes, and Fiona has 1 turtle. The graph shows the kind of animal each child owns, and how many he or she has.

- Oscar has an odd number of pets.
- Ali has 2 more pets than John has.
- Asia's only pet just had 7 babies!

Which pets do Oscar, Ali, and Asia own?

1 FIND OUT

What question do you have to answer?

_____

_____

Find out what the problem tells you.

2 CHOOSE A STRATEGY

I can _____ to solve the problem.

3 SOLVE IT

Use the graph and the clues.

Our Pets

		x			
		x			
		x			
		x			
		x		x	
	x	x		x	
x	x	x		x	x
x	x	x	x	x	x
Dogs	Cats	Fish	Turtles	Birds	Snakes

Which pets do Oscar, Ali, and Asia own?

Oscar _____

Ali _____

Asia _____

4 LOOK BACK

Read the problem again. Check your work.

Use or Make a Picture or Diagram

58

The students lined up their pencils to see whose was the longest. Now Marta, Alex, and Nia can't remember which pencils are theirs.

- Marta's is 1 inch longer than Ruben's.
- Alex's is the same length as Stella's.
- Nia's is 2 inches longer than Ray's.

How long are Marta, Alex, and Nia's pencils?

1 FIND OUT

What question do you have to answer?

_____

_____

Find out what the problem tells you.

2 CHOOSE A STRATEGY

I can _____ to solve the problem.

3 SOLVE IT

Use the graph and the clues.

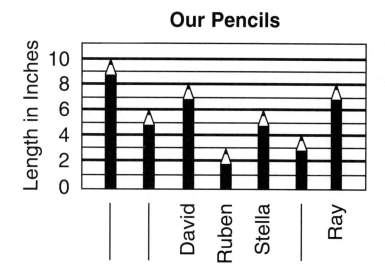

How long are Marta, Alex, and Nia's pencils?

Marta _____

Alex _____

Nia _____

4 LOOK BACK

Read the problem again. Check your work.

Thinking Questions
Questions to think about as you are solving problems

FIND OUT

What is the problem about?
What question do I have to answer?
What do I have to find out to solve the problem?
Are there any words or ideas I don't understand?
What information can I use?
Am I missing any information that I need?

CHOOSE A STRATEGY

Have I solved a problem like this before?
What strategy helped me solve it?
Can I use the same strategy for this problem?

SOLVE IT

What information should I start with?
Do I need to add or subtract?
How can I organize the information that I use or find?
Is the strategy I chose helpful?
Would another strategy be better?
Do I need to use more than one strategy?
Is my work easy to read and understand? Is it complete?

LOOK BACK

Did I answer the question that was asked in the problem?
Is more than one answer possible?
Is my math correct?
Does my answer make sense?
Can I explain why I think my answer is correct?